OCEANS AT RISK

Exploring Wonders, Facing Challenges,
Shaping an Ocean-Responsible Future

Zahra Jonsson

Cod ISBN: 9798867800086

Cover design by: Art Painter
Library of Congress Control Number: 2018675309
Printed in the United States of America

To those with curious eyes and open hearts, to you, adventurous readers who embrace the sea of knowledge with a desire for understanding and a passion for conservation. These pages are dedicated to you, guardians of our planet, because every word written here is a hymn to the beauty of the oceans and a cry of alarm for their vulnerability.

To those who, with the sound of waves in their ears, have dreamed of exploring the uncharted depths of knowledge. This book is a key to open doors hidden in the intricacies of the ocean of knowledge, an invitation to fearlessly dive into the depths of marine wonders.

I dedicate these pages to those who feel connected to the oceans, who consider them not only as vast expanses of water but as pulsating hearts of life, cradling secrets waiting to be unveiled. May this book ignite in you the flame of awareness and commitment, so that you become ambassadors of marine conservation.

To all those who believe in the power of small daily actions in shaping a sustainable future. May every word on these pages be a seed that sprouts in your mind and propels gestures that contribute to preserving the oceans for generations to come.

To those who dream of a world where the call of the oceans is an invitation to care, understand, and take action. May this book be your travel companion, guiding you through the currents of oceanic challenges and towards the shores of solutions and hopes.

Finally, this dedication is an acknowledgment of all those who commit themselves daily to the conservation of the oceans, who work tirelessly to protect these global treasures. May the words on these pages be an echo of your voices, amplifying your message of love for the oceans.

With gratitude and hope,
Zahra Jonsson

In the heart of the oceans resides ageless wisdom, a gentle summons inviting humanity to understand that conservation is not just a duty but a harmonious dance with life itself. We are custodians of this aquatic realm, called to protect and preserve, so that every wave tells tales of eternal beauty.

ZAHRA JONSSON

CONTENTS

INTRODUCTION

Welcome to a profound and reverent journey, a dive into the depths of the oceans that reveals both extraordinary beauty and heartbreaking vulnerability in one of the most crucial environments for life on Earth. In "Oceans at Risk," we will explore together the secrets whispered by ocean currents and confront the challenges that threaten the survival of these aquatic realms.

The Ocean: Vital Heart of the Earth

The vast and mysterious oceans are the heartbeat of our planet. They are the guardians of life, climate regulators, providers of oxygen, and reservoirs of biodiversity. In this introduction, we will delve into the vital heart of the Earth, exploring the profound connection that binds the oceans to every aspect of our existence. From microscopic organisms to immense whales, every oceanic creature contributes to an intricate balance that sustains our world.

An Exploration of Marine Wonders

Guided by scientists, scholars, and conservation enthusiasts, we will venture into the wonders of the oceans. Through the pages, we will encounter extraordinary creatures, explore kaleidoscopic coral reefs, and lose ourselves in the dark depths of the abyss. This section invites us to cast an intimate gaze on unknown worlds, captivating us with the surprising diversity that characterizes the oceans.

Ocean Challenges: Threats and Alarm Calls

But behind the serene facade of the waves, challenges lurk that demand our immediate attention. We will analyze marine pollution, from plastic littering beaches to chemicals poisoning the waters. We will confront overfishing, the privatization of marine resources, and climate change that is irreversibly transforming the oceans. Every word will be an alarm call, a call to action against the silent destruction threatening ocean life.

Shaping an Ocean-Responsible Future

Yet, we won't stop at the challenges; we will immerse ourselves in solutions. We will explore innovative ideas and sustainable practices that can change the course. From cutting-edge technology to global collaboration, each proposal is a step toward creating an ocean-responsible future. This section not only offers us hope but urges us to become active participants in ocean conservation.

The Invitation to Exploration

This book is an invitation to exploration, an opportunity to delve into the depths of the oceans without leaving the comfort of your reading space. I hope that every word plunges you into a fascinating journey and, at the same time, prompts you to reflect on how we can collectively defend the oceans from what threatens them.

In "Oceans at Risk," every page is a dive, every chapter is a wave carrying the power of awareness. I invite you to open these pages with open hearts and ready minds, prepared to explore, understand, and, above all, commit to the future of the oceans.

With gratitude for your interest and hope for our journey together,
Zahra Jonsson

PREFACE

Welcome to this journey through the uncharted depths of the oceans, a voyage that embraces wonder, confronts challenges, and will shape a future that is ocean-responsible. As I embark on this written exploration, I would like to share with you the pulsating heart behind every page of "Oceans at Risk."

The oceans are the lifeblood of the Earth, a symphony of life that has inspired poets, artists, and scientists for centuries. In the depths of these waters, age-old secrets lie hidden, creating a union between the majesty of nature and the fragility of our responsibility to the planet. I have written this book with the intent to share this profound connection and inspire a global commitment to marine conservation.

The exploration begins with admiration for the oceanic wonders. Through pages rich in detail, I invite you to immerse yourself in vibrant marine ecosystems, contemplate the astonishing variety of creatures dancing beneath the surface, and discover the magic that makes the oceans unique and indispensable for life on Earth.

However, this journey is not just a celebration of marine beauties. With critical eyes, we will examine the challenges endangering the oceans, threats ranging from insidious pollution to overfishing, from devastating climate change to the privatization of marine resources. Each challenge is an alarm call, an urgent invitation to change that must begin now.

The vision of a future that is ocean-responsible emerges through innovative and practical solutions. From engaging local communities to promoting cutting-edge technologies, we will explore concrete approaches to mitigate negative impacts and create a sustainable balance between humans and the sea.

This book is not just a collection of information; it is an invitation to action. Marine conservation is not a task for a few but a shared commitment for all who love this planet. I encourage you to become part of this mission, to translate words into actions that will shape the destiny of the oceans.

In "Oceans at Risk," I address you as allies in the guardianship of this global treasure. Through the pages of this book, I hope to convey the beauty, complexity, and urgency of marine conservation. We are all called to be custodians of the oceans, to join in a collective commitment to ensure that the waves continue to tell stories of life, hope, and wonder.

The author generated this text in part with GPT-3, OpenAI's large-scale language generation model. After generating the draft language, the author reviewed, modified, and revised the language at their discretion, taking ultimate responsibility for the content of this publication.

With gratitude for your commitment and in the hope of an ocean-responsible future,
Zahra Jonsson

PROLOGUE

On a full moon night, while the waves whispered age-old secrets and the silvery reflection of the satellite illuminated the vastness of the ocean, I began to write "Oceans at Risk." This prologue is the gateway through which I invite every reader to dive with me into the depths of the unknown, to explore the mysteries and threats hidden behind the apparent majesty of the waters.

The Call of the Ocean

I have always been fascinated by the call of the ocean, a song that fades into the infinite horizon. But this melody is more than a sweet serenade; it is a call to action. I felt the need to respond, to translate the waves into words, to give voice to the unknown inhabitants of the depths and to share their silent cry.

A Journey Among the Waves of Wonders

In the prologue, I invite you to step into an extraordinary journey. From the crystal-clear waters of the tropics to the furious storms of the Arctic, we will delve into worlds that defy imagination. Each page is an opportunity to explore extraordinary creatures, kaleidoscopic coral reefs, and the wonders hidden in the abyss.

Shadows Beneath the Surface

But beneath the serene surface, shadows lurk. In the prologue, we will become aware of the challenges that threaten the oceans. We will begin to unravel the plots of invisible pollution, overfishing that impoverishes ecosystems, and climate changes that alter the destiny of the oceans. Each wave carries a story of threat, a warning sign that we can no longer ignore.

An Appeal to Awareness and Action

The prologue is an appeal to awareness and action. I invite you to join me in recognizing that the beauty of the oceans is fragile, and our responsibility towards them is unavoidable. We can no longer afford to remain unaware of what is happening beneath the surface. Each story told is an invitation to become conscious guardians of this precious realm.

Visions of an Ocean-Responsible Future

But we don't stop at the challenges; the prologue introduces visions of an ocean-responsible future. We will explore innovative solutions, sustainable strategies, and the power of global collaboration. Each word is a step towards creating a world where the oceans are preserved for generations to come.

The Invitation to Journey with Me

This prologue is the invitation to journey with me, to navigate through the pages of "Oceans at Risk" with open eyes and committed hearts. I hope each word is like a tide that envelops you, transporting you to new worlds and, at the same time, urging you to become defenders of the oceans.

With deep gratitude for your interest and the hope that this journey together is a positive influence on the oceans and our future,
Zahra Jonsson

CHAPTER 1: THE OCEAN, VITAL HEART OF THE EARTH

The Ocean, with its vast blue expanse stretching to the horizon, is the silent guardian of life on Earth. When we gaze at the majestic ocean and admire its beauty, we often fail to fully grasp its complexity and crucial importance for the balance of our planet.

The Ocean: A World of Beauty and Mystery

Oceans are true unexplored worlds, underwater realms that capture the imagination and inspire wonder. The ocean depths conceal extraordinary creatures, from tiny luminescent organisms to majestic whales gracefully navigating the waters. The variety of colors and sounds animating the oceans creates an unparalleled visual and acoustic spectacle.

The beauty of the oceans is not only superficial; it is also hidden in the intricate networks of life that unfold beneath the surface. Colorful corals form underwater gardens that serve as a refuge and nursery for numerous species of fish and invertebrates. The intricate oceanic ecosystems are interconnected in a delicate balance, and every component, no matter how small it may seem, plays a crucial role in the tapestry of marine life.

The Vital Importance of Oceans

Oceans are not just beauty to behold; they are also essential for our survival. Every second, oceans produce a significant amount of oxygen thanks to photosynthetic microorganisms in their waters. About 50% of the oxygen, we breathe comes directly from the ocean, making the sea the vital lung of the Earth.

But the importance of oceans goes beyond oxygen production. Ocean currents act as gigantic thermal distribution systems, regulating the global climate and influencing weather patterns. Oceans also absorb large amounts of carbon dioxide, helping to mitigate the effects of climate change.

Biodiversity: Hidden Treasure of the Oceans

The richness of biodiversity in the oceans is a testament to the extraordinary variety of life they support. From tiny plankton forming the basis of the food chain to imposing predators dominating the depths, each creature has a unique role in the oceanic ecosystem.

Corals, in particular, are the forests of the oceans, providing shelter and nourishment to many marine species. The loss of these delicate ecosystems would have devastating impacts on the entire balance of the oceans and life on Earth.

Preserving the Ocean for Future Generations

It is our duty to preserve this wonder for future generations. The threats facing the oceans require concrete and conscious actions. Reducing pollution, protecting vulnerable marine areas, and promoting sustainable fishing practices are just some of the many challenges we must face.

Investing in marine scientific research is essential to fully understand the oceans and develop effective conservation strategies. Public education on the importance of the oceans plays a crucial role in engaging society in safeguarding this precious ecosystem.

In conclusion, the oceans are the vital heart of the Earth, pulsating with life and mystery. Exploring their beauty and understanding their importance is the first step towards creating a future where the oceans thrive and continue to support life on Earth.

CHAPTER 2: MARINE POLLUTION - A SILENT ENEMY

Oceans, once considered inexhaustible and immune, now face a silent yet pervasive threat: pollution. This chapter aims to explore the various forms of marine pollution, from plastic to chemical contamination, highlighting the challenges that arise and presenting innovative solutions to preserve the health of the oceans.

Plastic Pollution: Oceans Saturated with Plastic

The most evident enemy plaguing the oceans is represented by plastic. Millions of tons of plastic waste pour into the seas every year, forming floating islands that irreparably damage marine ecosystems. Marine creatures, deceived by the resemblance to food, ingest plastic, causing internal damage and death.

Addressing this problem requires a multifaceted approach. Reducing the use of single-use plastic through stricter laws and awareness campaigns is essential. Simultaneously, innovation in recycling and the development of biodegradable materials can reduce the impact of plastic on the marine environment.

Chemical Contamination: Poison in the Seas

In addition to plastic, oceans are threatened by a wide range of harmful chemicals. From the chemical industry to agricultural activities, contaminants reach the oceans through waterways and atmospheric precipitation, poisoning marine life and compromising human food safety.

To address chemical contamination, it is essential to regulate industrial and agricultural emissions, implement sustainable farming techniques, and invest in water purification. The adoption of advanced technologies, such as green filters to absorb contaminants, could represent a breakthrough in removing harmful chemicals from the oceans.

Microplastics: Tiny Invisible Killers

A frequently overlooked enemy is represented by microplastics, tiny fragments resulting from the degradation of larger objects. These microscopic particles infiltrate all layers of the marine ecosystem, damaging organisms of every size. The food chain, including humans, is at risk.

To address microplastics, a global approach is necessary. Reducing the use of plastic at the source, improving wastewater treatment systems, and developing advanced cleaning techniques are all crucial strategies to eliminate this invisible threat.

Innovative Solutions: Preserving Oceans for Future Generations

To counter marine pollution, innovative solutions and global joint efforts are required. The use of advanced technologies, such as ocean drones equipped with sensors to monitor and collect pollution data, can improve our understanding of the problem and guide targeted actions.

Furthermore, encouraging the adoption of sustainable packaging and developing biodegradable materials can significantly reduce the presence of plastic in the oceans. Educational campaigns on proper waste disposal and awareness of marine pollution are fundamental to engage society in the fight against this silent enemy.

Conclusions: A Future Without Marine Pollution

In conclusion, marine pollution is an urgent threat that

requires immediate and sustained actions. With a combination of stricter legislation, technological innovations, and a cultural shift towards more sustainable practices, we can reverse the trend and preserve the beauty and vitality of the oceans for future generations. Addressing marine pollution is not just an environmental necessity but a moral imperative to protect the vital heart of the Earth.

CHAPTER 3: OCEAN GRABBING - WHEN THE SEA BECOMES PRIVATE PROPERTY

Oceans, once considered vast and unexplored territories, are now at the center of a growing threat: ocean grabbing. This chapter will explore this phenomenon, where marine resources are taken away from local communities, analyzing its roots, impacts, and proposing legal and social strategies to counteract this privatization and ensure fair access to oceanic resources.

The Ocean as a Common Good: History and Tradition

For centuries, oceans have been considered common goods, sources of sustenance and resources for coastal communities. Artisanal fishing and the collection of seafood were sustainable activities integrated into the natural cycles of marine ecosystems. However, over time and with technological development, perspectives on ocean use have changed.

Roots of Ocean Grabbing

Ocean grabbing has its roots in a combination of factors, including the increasing global demand for marine resources, advanced technology in industrial fishing, and the lack of effective regulations.

Large corporations and industrialized nations, driven by the desire for profit, began to see oceans as a source of wealth to exploit without considering the impacts on local communities dependent on these resources.

Impacts of Ocean Grabbing on Local Communities

The phenomenon of ocean grabbing has devastating effects on local communities dependent on fishing and marine resources. The privatization of waters limits traditional communities' access to fishing areas, leading to a loss of livelihoods and threatening their traditions and ways of life. The competition for fishing resources becomes increasingly fierce, with large corporations often having superior financial and technological resources, leaving local communities at a disadvantage.

Legally Binding Strategies: Protecting Local Interests

To counter ocean grabbing, it is essential to implement legally binding strategies at both international and national levels. The creation and strengthening of laws protecting the rights of local communities to access oceanic resources are crucial. Additionally, exclusive fishing zones for traditional communities must be established, ensuring they have priority in the sustainable use of marine resources in their territorial waters.

Community Participation and Awareness

Beyond laws, actively involving local communities in decisions regarding the use of marine resources is crucial. Empowering communities through educational programs and promoting community participation in decision-making processes can reduce the vulnerability of local communities in the face of ocean grabbing.

Participatory Management Models: The Future of Marine Conservation

A promising approach to counteract ocean grabbing is represented by participatory management models. Involving local communities, researchers, and government authorities in defining strategies for the management of marine resources can create a balance between sustainable exploitation and the conservation of marine ecosystems. These models can be adapted to the specificities of each region, taking into account the needs of local communities and promoting the sustainable use of marine resources.

Monitoring and Accountability of Businesses

Another crucial tactic is actively monitoring business activities in the marine environment. Implementing tracking systems and transparent publication of business practices can help identify and stop companies attempting to illegally exploit marine resources.

Global Responsibility: A Shared Commitment

Addressing ocean grabbing requires a shared commitment globally. Nations must collaborate to establish international norms and regulations that protect the oceans as a common heritage of humanity. Furthermore, it is essential to raise public awareness about the seriousness of the problem and the need to protect local communities and marine ecosystems.

Conclusions: Safeguarding the Heart of the Oceans

In conclusion, ocean grabbing represents a direct threat to local communities and the health of the oceans. Through an integrated approach that includes binding laws, community participation, and corporate responsibility, we can protect the vital heart of the oceans. Only with global commitment and widespread awareness can we ensure that the oceans remain common heritage, preserving their beauty and richness for future generations.

CHAPTER 4: SUSTAINABLE AQUACULTURE - CULTIVATING THE FUTURE OF THE SEA

Aquaculture, the practice of cultivating aquatic organisms in controlled environments, emerges as a key solution to meet the growing global demand for seafood without depleting marine resources. In this chapter, we will explore the pros and cons of aquaculture and present innovative models aimed at ensuring sustainable food production without compromising the health of the oceans.

Aquaculture in Perspective

Aquaculture has experienced exponential growth in recent decades, becoming a significant source of fish for human consumption. However, its rapid expansion has raised some concerns, including negative environmental impacts, animal welfare issues, and challenges related to food quality and safety.

Pros of Aquaculture

Response to Global Fish Demand: With traditional fishing reaching its sustainable limits, aquaculture offers an efficient way to meet the growing demand for fish and seafood without excessively exploiting natural marine resources.

Job Creation: Aquaculture operations can create job opportunities in coastal communities, contributing to supporting the local economy and reducing dependence on traditional fishing.

Environmental Control: Unlike open-sea fishing, aquaculture provides greater control over the growth environment of marine organisms, allowing for more targeted management and reducing the risk of overfishing.

Cons of Aquaculture

Local Environmental Impacts: The accumulation of organic waste, chemicals, and medications used in aquaculture can have negative impacts on the quality of surrounding water and local marine ecosystems.

Disease Spread: The concentration of organisms in confined spaces can promote the spread of diseases among aquatic populations, necessitating the use of antibiotics and chemicals that can contaminate the environment.

Exploitation of Food Resources: Some aquaculture practices require large quantities of wild fish for feed production, leading to a paradox where wild fish are harvested to feed cultured organisms.

Innovative Models of Sustainable Aquaculture

To address the challenges associated with aquaculture, innovative and sustainable approaches are needed to balance food production with the conservation of marine ecosystems. Some emerging models include:

Integrated Multi-Trophic Aquaculture (IMTA): This approach involves the simultaneous cultivation of diverse species in a single environment. For example, algae can be cultivated alongside fish, creating a system where nutrients released by one species benefit others.

Closed-Loop Aquaculture: This model involves the use of closed systems where water is recycled and treated to reduce environmental impact. This reduces the need to extract water from the sea and minimizes the release of nutrients into the surrounding environment.

Development of Sustainable Feeds: Reducing dependence on wild fish for feed production is essential for sustainable aquaculture. Research focuses on producing feeds based on plants, algae, or insects, thereby reducing pressure on fishing.

Legal and Social Strategies for Sustainability

To ensure a sustainable future for aquaculture, it is essential to integrate legal and social approaches that promote responsible management of marine resources.

Environmental Regulations: Laws and regulations must be implemented and strengthened to ensure environmentally respectful aquaculture practices. These regulations should cover aspects such as the use of antibiotics, waste management, and water resource conservation.

Community Engagement: Actively involving local communities in the planning and implementation of aquaculture operations is essential. This can promote more responsible and sustainable management of marine resources while preserving the traditional way of life of coastal communities.

Sustainability Certifications: Encouraging the adoption of sustainability standards and certifications can guide companies towards more responsible practices. For example, the Marine Stewardship Council (MSC) certifies sustainable fishing, while the Aquaculture Stewardship Council (ASC) focuses on sustainable aquaculture.

The Vision of a Fair Future

Sustainable aquaculture represents a promising way to cultivate the future of the sea without compromising the oceans. Leveraging technological innovations, advanced management practices, and engaging local communities is crucial to create a balance between food production and the conservation of marine ecosystems.

In conclusion, investing in sustainable aquaculture models is an urgent necessity to preserve marine biodiversity and ensure that marine resources are available for future generations. Only through a holistic approach and collaboration between governments, businesses, and local communities can we cultivate the future of the sea sustainably, ensuring that the oceans continue to thrive as a source of life and nourishment.

CHAPTER 5: OVERFISHING - A THREAT TO MARINE BALANCES

The shadow of overfishing looms as an imminent danger over the oceans, threatening marine balances and the survival of numerous species. In this chapter, we will explore the devastating effects of overfishing on marine life and suggest approaches based on eco-sustainable fisheries management to preserve marine ecosystems.

Expansion and Effects of Overfishing

Overfishing is the result of excessive fishing compared to the natural reproductive capacity of fish populations. This practice, fueled by the growing global demand for seafood, has significant impacts on marine ecosystems.

Decline of Fish Populations: Overfishing pushes many fish populations to collapse, threatening marine biodiversity and compromising ecosystem stability.

Trophic Imbalances: The excessive removal of key predators, such as tuna and sharks, can trigger trophic imbalances, with an increase in prey populations and a decrease in the diversity and resilience of ecosystems.

Impact on Coastal Communities: Coastal communities, often dependent on fishing for their livelihoods, suffer severe economic and social consequences due to the collapse of fish resources.

Approaches to Overfishing

Addressing overfishing requires a multifaceted approach that includes sustainable management strategies, conservation measures, and actions to engage local communities.

Science-Based Management: Fisheries management should be based on accurate scientific data. Monitoring fish populations, assessing the carrying capacity of marine ecosystems, and establishing sustainable fishing quotas are crucial steps to avoid overfishing.

Fishing Quotas: Implementing limited fishing quotas helps ensure that fish extraction does not exceed the natural reproductive capacity of populations, allowing them to maintain a sustainable presence in the oceans.

Marine Reserves: Creating marine reserves, areas where fishing is prohibited or restricted, provides a safe refuge for declining fish populations. These reserves act as biodiversity reservoirs, contributing to maintaining trophic balances.

Innovative Technologies in Overfishing Management

The use of innovative technologies can significantly enhance efforts to manage overfishing and restore marine balances.

Remote Monitoring Technologies: Drones and satellites can be used to monitor fishing activities, detect illegal practices, and ensure compliance with regulations.

Supply Chain Tracking Systems: Implementing tracking systems can ensure that the caught fish comes from sustainable sources. This provides transparency in the supply chain and enables consumers to make informed decisions.

Innovations in Selective Fishing: Developing selective fishing techniques can reduce bycatch, the accidental capture of non-target species. More selective nets and adapted fishing practices can help preserve marine diversity.

Engaging Local Communities in Sustainable Management

Active engagement of local communities is essential to implementing sustainable fisheries management strategies and preserving marine resources.

Resource Co-Management: Involving local communities in co-managing fish resources promotes more responsible and sustainable management. Local knowledge can significantly contribute to identifying effective practices and understanding ecosystem changes.

Support for Economic Alternatives: Creating economic alternatives for communities closely tied to fishing can reduce pressure on overfishing. Sustainable development programs can encourage economic diversification in these regions.

Education and Awareness: Educational programs and awareness initiatives can inform local communities about the sustainability of fishing and ways to preserve fish resources for future generations.

The Need for Global Cooperation

Overfishing is a challenge that transcends national borders, requiring global cooperation to address its causes and consequences.

Integration of Global Policies: Global policies must be integrated to ensure that conservation efforts are consistent internationally. Agreements such as the United Nations Convention on the Law of the Sea provide a foundation but need to be strengthened and more effectively implemented.

Combatting Illegal, Unreported, and Unregulated (IUU) Fishing: IUU fishing significantly contributes to overfishing. International cooperation is essential to counter this practice, through information sharing and the adoption of punitive measures against states and companies involved.

Conclusion: Safeguarding Marine Ecosystems

Overfishing represents a direct threat to marine life and the sustainability of marine ecosystems. Only through holistic, science-based, innovative, and socially responsible management can we preserve fish resources and ensure that the oceans maintain their richness and diversity. The future of marine ecosystems depends on our ability to adopt sustainable fishing practices and cooperate globally to protect the hidden treasures of the oceans.

CHAPTER 6: PIRATE FISHING - THE CRIME OF THE ABYSS

In the vast theater of the oceans, an obscure drama unfolds pirate fishing. This chapter will delve into the phenomenon of pirate fishing, examining its roots, the devastating impacts on vulnerable marine species, and providing innovative strategies to combat this marine crime and ensure the security of global waters.

The Mystery of Pirate Fishing

Pirate fishing is a term that conjures images of mysterious ships moving in the darkness of the oceans, silently plundering marine resources without regard for laws or regulations. This phenomenon, often linked to illegal, unreported, and unregulated (IUU) fishing, threatens marine balances and jeopardizes the sustainability of fishery resources.

The Roots of the Problem

Pirate fishing has deep-seated roots, often tied to economic motivations and the lack of effective regulations. Some of the main factors fueling this marine crime include:

Lack of Control and Surveillance: The vast expanses of the oceans make it challenging for authorities to monitor and control fishing activities. This creates a favorable environment for illegal activities.

Short-Term Profit: Enterprises involved in pirate fishing are often driven by the desire for short-term profit, without concern for the long-term consequences on the sustainability of fishery resources.

Weakness of Laws and Sanctions: In many regions, laws against illegal fishing may be weak or difficult to enforce. The lack of significant sanctions can encourage illegal practices.

The Devastating Impacts of Pirate Fishing

Pirate fishing has devastating effects on marine ecosystems and coastal communities dependent on fishing.

Collapse of Fish Populations: Overfishing resulting from pirate fishing can lead to the collapse of fish populations, jeopardizing marine biodiversity and the food security of numerous communities.

Damage to Marine-Coastal Ecosystems: Destructive practices of pirate fishing, such as the use of trawl nets, can damage marine habitats and compromise the reproduction of many species.

Economic Losses for Local Communities: Coastal communities relying on fishing for their livelihoods suffer significant economic losses due to unfair competition and the decline of fish resources.

Innovative Strategies to Combat Pirate Fishing

Addressing pirate fishing requires innovative approaches that go beyond the mere application of laws and sanctions. Some key strategies include:

Advanced Surveillance Technologies: The use of advanced technologies, such as high-resolution satellites, marine drones, and ship-tracking systems, can significantly improve the surveillance of fishing activities and enable timely intervention against pirate fishing.

International Collaboration: Pirate fishing is a problem that crosses borders. International collaboration is essential to exchange information, coordinate actions, and implement global regulations against this practice.

Transparency in the Supply Chain: Making the supply chain of seafood more transparent can help identify and stop pirate fishing. The adoption of tracking systems and sustainability certifications can ensure that consumers are well-informed about companies' practices.

Strictly Enforced Penalties: Strengthening sanctions against pirate fishing is essential to deter this illegal activity. Significant penalties, asset confiscation, and fishing activity bans can serve as deterrents.

Involvement of Local Communities: Local communities must be actively involved in protecting fishery resources. Awareness programs and training initiatives can help communities recognize and report pirate fishing activities.

The Need for Global Commitment

To defeat pirate fishing, global commitment involving governments, international organizations, businesses, and individuals is necessary.

Strengthening International Regulations: International organizations must work together to establish more stringent global standards and regulations against pirate fishing. The United Nations Convention on the Law of the Sea provides a framework, but it needs to be strengthened and effectively enforced.

Monitoring Sensitive Areas: Identifying and monitoring the most sensitive marine areas, such as marine reserves and critical habitats, can reduce the vulnerability of these areas to pirate fishing.

Economic Incentives for Sustainable Fishing: Creating economic incentives for sustainable fishing can encourage businesses to adopt environmentally friendly practices and discourage pirate fishing.

Conclusion: Safeguarding the Oceans for Future Generations

Combating pirate fishing requires a joint global commitment to protect the oceans and preserve marine life. Only through innovation, international collaboration, and a shift in business practices can we put an end to this crime of the abyss and ensure a sustainable future for fishery resources and marine ecosystems. The challenge is significant, but the reward is the opportunity to keep the oceans alive and thriving for future generations.

CHAPTER 7: CLIMATE CHANGE - THE CRISIS EMBRACING THE OCEANS

The oceans, invaluable custodians of life on Earth, are at the center of a growing crisis: climate change. In this chapter, we will examine the devastating impacts of climate change on the oceans, including acidification and sea-level rise. We will propose revolutionary solutions to mitigate and adapt to these changes, preserving the stability of marine ecosystems.

Climate Change and Oceans: A Broken Bond

The oceans, traditionally shaped by natural forces, are now undergoing an unprecedented change due to human activities. The increase in greenhouse gas emissions is triggering a series of impacts that threaten the delicate balances of marine ecosystems.

Global Warming: The rise in atmospheric temperatures contributes to the warming of the oceans, causing global-scale impacts such as the weakening of coral reefs and the alteration of marine currents.

Ocean Acidification: The absorption of excessive amounts of atmospheric carbon dioxide by the oceans is causing the acidification of seawater. This phenomenon threatens organisms like corals, mollusks, and fish that depend on the formation of calcium carbonate.

Sea-Level Rise: Global warming leads to the melting of glaciers and polar ice caps, contributing to the rise in sea levels. This puts coastal communities and vital habitats for many marine species at risk.

Ocean Acidification: A Silent Threat

Ocean acidification represents a silent but potentially devastating threat to marine ecosystems.

Impacts on Corals: Corals, already threatened by bleaching due to warming, are particularly sensitive to acidification. The formation of their calcium carbonate skeleton may be compromised, jeopardizing the survival of coral reefs.

Mollusks and Gastropods: Marine organisms that rely on the formation of calcium carbonate shells and shells, such as mollusks and gastropods, are vulnerable to acidification. This threatens the marine food chain and the biodiversity of the oceans.

Impacts on the Food Chain: Organisms that make up the marine food chain, including fish and marine mammals, may experience negative effects of acidification. This can have cascading consequences on the entire food chain and the communities that depend on marine resources.

Sea-Level Rise: A Global Threat

Sea-level rise is a direct consequence of global warming and the melting of glaciers and polar ice caps.

Risks for Coastal Communities: Coastal communities are particularly vulnerable to sea-level rise. Densely populated areas, such as coastal cities and river deltas, are at risk of flooding and land loss.

Loss of Critical Habitats: Many marine species depend on specific habitats, such as mangroves and coastal wetlands. Sea-level rise threatens the stability of these habitats, jeopardizing the survival of numerous species.

Impacts on Navigation and Maritime Economy: Sea-level rise can influence navigation and maritime economy. Ports, infrastructure, and shipping routes may be compromised, with significant impacts on global trade.

Mitigation and Adaptation: A Crucial Dual Approach

To address the impacts of climate change on the oceans, a combined approach of mitigation and adaptation is necessary.

Mitigation through Renewable Energy: Reducing greenhouse gas emissions is essential to mitigate global warming and, consequently, the impacts on the oceans. Investing in renewable energies such as solar and wind can reduce dependence on fossil energy sources.

Marine Reserves and Habitat Conservation: Creating marine reserves and protecting critical habitats, such as mangroves and coastal wetlands, is crucial for the conservation of marine biodiversity and providing vital spaces where species can adapt to changes.

Research and Technological Innovation: Investing in research and technological innovation can lead to revolutionary solutions to address specific issues related to climate change. For example, technologies that promote the growth of corals resistant to acidification could offer hope for the survival of coral reefs.

Revolutionary Solutions: A Sustainable Future for the Oceans

Responsible Climate Engineering: Climate engineering, if conducted responsibly and based on thorough scientific and ethical assessments, could offer innovative solutions to mitigate the effects of climate change on the oceans. For example, stimulating oceanic photosynthesis could help absorb more carbon dioxide.

Carbon Capture and Storage (CCS) Technologies: CCS technologies can capture carbon dioxide directly from the atmosphere or emission sources and store it safely. This approach can reduce greenhouse gas concentrations and mitigate ocean acidification.

Marine Restoration Projects: Marine restoration initiatives, such as planting mangroves and creating oyster reefs, can contribute to restoring marine habitats and increasing ecosystem resilience to climate impacts.

Global Collaboration as the Key to Success

Addressing the climate crisis embracing the oceans requires global collaboration and a joint commitment to adopting innovative and sustainable solutions.

Effective International Agreements: International agreements must be strengthened and implemented to ensure the reduction of global emissions and the conservation of the oceans. Collaboration between nations is essential to address this challenge on a global scale.

Education and Awareness: Public awareness of the importance of the oceans and the risks of climate change is crucial. Education can inspire individual and collective actions to reduce the impact on the marine environment.

Economic and Social Sustainability: Integrating sustainable practices into the economy and society is crucial to ensuring a future where the oceans can thrive. This requires a shift in collective mindset and a commitment to sustainable development models.

Conclusion: Safeguarding the Blue Heart of the Earth

The oceans, the blue heart of the Earth, are in danger due to climate change. However, with bold actions, scientific innovations, and unprecedented global collaboration, it is possible to preserve the beauty and vitality of the oceans for future generations. Each of us has a crucial role to play in shaping a sustainable future for the oceans and the planet as a whole. It is time to act with respect and gratitude toward our precious marine environment.

CHAPTER 8: MARINE BIODIVERSITY - TREASURES TO SAFEGUARD

The oceans, vast and mysterious, are the stage for an unparalleled wealth: marine biodiversity. In this chapter, we will delve into the extraordinary diversity of life that animates the ocean depths and illustrate innovative approaches to protect and preserve endangered marine species.

The Ballet of Ocean Life

The oceans are the venue for an unparalleled spectacle, a captivating ballet in which marine species play unique and interconnected roles. From microscopic planktonic creatures to majestic marine mammals, every form of life contributes to creating a delicate balance within marine ecosystems.

The Riches of Coral: Coral reefs, true underwater cities, harbor an extraordinary variety of species. From vibrant tropical fish to delicate anemones, biodiversity in coral areas is a treasure to be preserved.

The Epic Migration of Marine Animals: From whales traversing the oceans in long migrations to marine turtles navigating through the seas to lay eggs, the stories of marine animal migration are epic and crucial for biodiversity.

Microcosm of Plankton: Even microscopic planktonic

organisms are essential for the marine food chain. From phytoplankton producing oxygen through photosynthesis to zooplankton nourishing many marine species, this microcosm plays a vital role.

Threats to Marine Biodiversity

Despite the richness of marine life, the biodiversity of the oceans is threatened by multiple challenges.

Climate Change: Global warming and ocean acidification jeopardize marine habitats and threaten the survival of many species, especially those linked to corals and calcium carbonate shells.

Marine Pollution: From the accumulation of plastic to toxic chemical substances, marine pollution poses a direct threat to many marine species. Turtles ingesting plastic waste and poisoned marine mammals are just some examples of its effects.

Overfishing: Overfishing jeopardizes the survival of many fish species and disrupts the trophic balances of marine ecosystems. Overfishing can lead to the collapse of fish populations, with serious consequences for biodiversity.

Innovative Approaches for Conservation

Preserving marine biodiversity requires a joint commitment to adopting innovative and sustainable approaches.

Advanced Monitoring Technologies: Utilizing advanced technologies such as marine drones and satellites to monitor marine populations and critical habitats. These tools provide crucial data for management and conservation.

Protection of Marine Areas: Creating marine reserves and no-fishing zones is an effective way to protect critical habitats and provide safe spaces for the reproduction and growth of marine species.

Reintroduction of Marine Species: In some cases, the reintroduction of threatened marine species can be an effective strategy to restore ecological balances. This requires careful management and a deep understanding of ecosystems.

Marine Habitat Restoration: Marine restoration initiatives, such as planting mangroves and creating oyster beds, can contribute to restoring vital habitats for many marine species.

Successful Conservation Projects

Many conservation projects have shown that commitment and innovation can make a difference in protecting marine biodiversity.

Marine Turtle Conservation Project: In many regions, conservation projects work to protect marine turtles and their habitats. The use of red lights on beaches during egg deposition reduces misorientation of hatchlings, increasing their chances of survival.

Environmental Awareness Programs: Environmental awareness initiatives aimed at educating local communities and actively involving them in marine biodiversity protection. These programs promote awareness and respect for the oceans.

Successful Marine Reserves: Well-managed marine reserves, such as the Great Barrier Reef Marine Park in Australia, have proven effective in protecting biodiversity and restoring marine ecosystems.

Convergence between Conservation and Communities

Engaging local communities in marine biodiversity conservation is crucial for long-term success.

Resource Co-Management: Involving local communities in the co-management of marine resources promotes more responsible and sustainable management. Communities become active partners in ocean conservation.

Support for Sustainable Local Economies: Creating sustainable economic opportunities, such as ecotourism and sustainable fishing, can reduce pressure on marine resources and promote conservation.

Environmental Education in Schools: Introducing environmental education programs in schools to raise awareness among new generations about the importance of marine biodiversity and the role each person can play in its conservation.

The Need for Global Commitment

Conserving marine biodiversity is a challenge that goes beyond national borders. Global commitment is required to address threats to marine life and ensure a sustainable future for the oceans.

International Collaboration: Collaboration between nations is essential to address transnational threats to marine biodiversity. International agreements must be strengthened and effectively implemented.

Reduction of Greenhouse Gas Emissions: Addressing climate change is crucial to protect marine biodiversity. Reducing greenhouse gas emissions through the adoption of sustainable practices and the use of renewable energy is crucial.

Global Awareness: Promoting global awareness of the importance of marine biodiversity through media campaigns, events, and online initiatives. Public awareness can influence behaviors and decisions at both individual and collective levels.

Conclusion: Guardians of the Oceans

Marine biodiversity is the hidden treasure of the oceans, a kaleidoscope of life that depends on our care and dedication. Protecting this wealth requires unwavering commitment, uniting the forces of local communities, scientific institutions, businesses, and governments worldwide. We are the guardians of the oceans, and only through global commitment can we ensure that this extraordinary spectacle of life continues to enchant future generations. The road is challenging, but the reward is invaluable: oceans rich in life and treasures to discover for generations to come.

CHAPTER 9: GREEN TECHNOLOGIES - INNOVATIONS FOR MARINE CONSERVATION

In the vast realm of the oceans, technology proves to be a powerful ally in the fight for marine conservation. In this chapter, we will explore cutting-edge technologies, such as oceanic drones and artificial intelligence, that can be employed to monitor and protect the oceans in new and efficient ways.

The Power of Green Technologies

Green technologies offer a new level of precision and efficiency in monitoring and conserving the oceans. These innovations promise to transform how we understand and preserve marine ecosystems.

Oceanic Drones: Eyes in the Deep Abyss

Oceanic drones represent a revolution in our ability to explore and monitor the oceans. These automated vessels, often inspired by the form of marine creatures, are equipped with advanced sensors and can dive to depths inaccessible to humans.

Pollution Monitoring: Oceanic drones can patrol oceanic areas in search of accumulations of plastic, toxic chemicals, and other forms of pollution. Real-time data collection provides a detailed map of pollution impacts.

Detection of Marine Species: Using advanced detection techniques, drones can locate and monitor marine species, providing crucial information for conservation. This is particularly useful for elusive or threatened species.

Exploration of Marine Habitats: Drones can explore marine habitats, such as coral reefs and submarine canyons, documenting their health and identifying any signs of degradation or changes.

Artificial Intelligence (AI): Intelligent Data Analysis

Artificial intelligence plays an increasingly important role in the management and analysis of oceanic data. This technology can process vast amounts of information quickly and efficiently, providing insights that would otherwise be challenging to obtain.

Detection of Climate Change: AI can analyze historical and real-time data to identify patterns and trends related to climate change. This helps predict and mitigate future impacts on marine communities.

Anti-Pirate Fishing Surveillance: The use of advanced algorithms can identify suspicious behavioral patterns associated with pirate fishing. AI can analyze satellite and ship tracking data to detect illegal activities.

Aquaculture Monitoring: In aquaculture, AI can optimize cultivation practices, monitor fish health, and prevent potential epidemics. This contributes to making aquaculture more sustainable and resilient.

Technologies for Marine Conservation

Species Tracking Systems: Using advanced tracking technologies, such as satellite transponders, it is possible to monitor the movements of marine species. This is crucial for the management of fish populations and the conservation of threatened species.

Synthetic Biology for Assisted Reproduction of Species: Synthetic biology offers revolutionary possibilities in the assisted reproduction of threatened marine species. This technology could be used to strengthen the genetics of weak populations and increase genetic diversity.

Smart Environmental Sensors: Intelligent sensors can be deployed in the oceans to monitor key environmental parameters such as temperature, acidity, and oxygen levels. This data provides a detailed picture of ocean health and helps identify any anomalies.

Plastic Removal Technologies: Solar-powered floating devices and floating barriers are some of the proposed technologies to collect and remove plastic waste from the oceans.

Successes and Challenges of Green Technologies

Successes in Fishing Surveillance: The use of technologies such as satellite tracking and AI-based surveillance systems has led to successes in the fight against illegal fishing and overfishing. Joint efforts by organizations, governments, and businesses have helped identify and stop ships involved in illegal activities.

Surveillance of Marine Reserves: Advanced monitoring technologies are improving the surveillance of marine reserves, ensuring that no-fishing rules are respected and protecting crucial habitats.

Challenges in Widespread Adoption: Despite successes, the widespread adoption of some technologies is still limited by factors such as high costs, lack of infrastructure, and regulatory issues. Overcoming these challenges is crucial to maximize the impact of green technologies.

The Ethics of Green Technology

The use of green technologies in marine conservation also raises fundamental ethical questions.

Privacy and Surveillance: Monitoring through drones and surveillance systems can raise concerns about the privacy of individuals and local communities. It is essential to develop policies and regulations that balance the need for surveillance with respect for privacy.

Socioeconomic Impact: The introduction of advanced technologies can influence local communities, especially those traditionally dependent on fishing. An ethical approach requires consideration of socioeconomic impacts and the creation of inclusive solutions.

Global Accessibility: Ensuring that technologies are accessible globally is essential to address marine challenges on a worldwide scale. This requires international collaborations and concrete commitment to overcome economic and technological divides.

The Future of Green Technologies for the Oceans

The future of green technologies for marine conservation is promising but requires ongoing commitment and global collaborations.

Investments in Research and Development: Greater investments in research and development are needed to develop more efficient, accessible, and sustainable technologies for marine conservation.

International Collaboration: Collaboration between nations is essential to address marine challenges on a global scale. Sharing knowledge, data, and resources can accelerate the development and adoption of green technologies.

Training and Awareness: Training and awareness are crucial to ensure that local communities and involved stakeholders are well-informed about the ethical and responsible use of green technologies.

Conclusion: Technologies for a Blue Future

Green technologies represent a crucial key to open the doors to a sustainable future for the oceans. With a combination of technological innovation, global commitment, and ethical vision, we can protect and preserve the oceans for future generations. Every step forward, whether small or large, brings us closer to a blue future where oceans thrive, and marine life continues to inspire and amaze.

CHAPTER 10: COASTAL COMMUNITIES - GUARDIANS OF THE OCEAN

In the pursuit of marine conservation, the focus often turns to large global efforts, but nearby the waves and shores, coastal communities stand as guardians of the ocean. In this chapter, let's delve into the stories of coastal communities committed to marine conservation and explore participatory management models that actively involve people in safeguarding the oceans.

Deep Bonds with the Ocean

Coastal communities have an intrinsic connection with the ocean. The waves are the soundtrack of their days, and marine life is woven into the fabric of their lives. These deep bonds create a natural responsibility to protect and preserve the oceans for future generations.

Local Success Stories: From small fishing villages to coastal cities, communities worldwide are embracing the challenge of conserving the marine environment. For instance, the community of Cabo Pulmo in Mexico transformed an overfished area into a thriving marine reserve through participatory management.

Sustainable Traditional Fishing: In many coastal communities, traditional fishing techniques are passed down from generation to generation. These practices often include selective fishing and the use of local knowledge to ensure sustainable management of fish resources.

Culture of Respect for the Ocean: In many coastal cultures, there exists a strong culture of respect for the ocean. Festivals and celebrations often reflect gratitude for the gifts of the ocean and a commitment to preserving its beauty.

Participatory Management Models

Actively involving communities in ocean management is essential for the long-term success of marine conservation. Various models demonstrate that collaboration between local communities, non-governmental organizations, and government authorities can lead to significant outcomes.

Community Marine Reserves: Establishing marine reserves managed by the communities themselves is an effective conservation model. Communities define rules and practices that promote sustainability and conservation of marine habitats.

Cooperative Fishing: Forming fishers' cooperatives enables more sustainable management of fish resources. Communities collectively decide on fishing quotas and implement practices that safeguard the diversity of marine species.

Recycling and Waste Reduction Programs: Involving communities in coastal recycling and waste reduction programs helps prevent marine pollution. Initiatives like beach cleanups and recycling awareness actively engage residents in safeguarding the marine environment.

Success Stories of Coastal Communities

Cabo Pulmo, Mexico: The Cabo Pulmo community transformed an overfished area into a successful marine reserve. Actively involving local fishermen in management, the reserve has witnessed a remarkable recovery of biodiversity and created economic opportunities through sustainable tourism.

Participatory Management in the Philippines: In many

coastal communities in the Philippines, participatory management projects have led to the regeneration of coral habitats and the restoration of fish populations. Collaboration between fishermen, environmental organizations, and local authorities played a key role in this success.

Awareness Programs in Thailand: In some coastal locations in Thailand, awareness programs have led to increased awareness of the importance of marine conservation. This has stimulated the adoption of sustainable practices and active participation in ocean protection.

Challenges and Opportunities

Economic Pressures: Coastal communities often face economic pressures that can make it challenging to adopt sustainable practices. Creating alternative economic opportunities, such as sustainable tourism, can help mitigate these challenges.

Climate Change and Adaptive Responses: Coastal communities must address the impacts of climate change, including rising sea levels and extreme weather events. Developing adaptive and resilient responses is crucial.

Access to Resources and Aid: Ensuring that coastal communities have fair access to resources and receive aid when needed is crucial to support conservation efforts. Equity is a key element to ensure ongoing participation and collaboration.

Engaging New Generations

Engaging new generations is crucial to ensure the continuity of coastal community conservation efforts.

Environmental Education in Schools: Integrating environmental education programs into schools in coastal communities helps instill awareness of the importance of marine conservation from a young age.

Participation Opportunities for Youth: Creating opportunities for young people to actively participate in marine conservation, through volunteer programs, local research initiatives, and specific projects.

Inclusion of Youth Voices in Decisions: Involving youth in decisions about marine management ensures a fresh and innovative perspective, promoting a sense of responsibility for the future.

The Strength of Coastal Communities

Coastal communities represent a fundamental force in ocean conservation. Their intimate connection with the marine environment, the history of local successes, and commitment to participatory management demonstrate that when communities take the lead in safeguarding the oceans, the results can be extraordinary. The ocean is their treasure, and as dedicated guardians, coastal communities are writing significant chapters in the history of marine conservation. Through ongoing collaboration and global commitment, we can ensure that these success stories multiply, making coastal community's true custodians of the ocean for generations to come.

CHAPTER 11: MARINE ECO-TOURISM - A SUSTAINABLE PATH TO EXPLORE THE OCEANS

Marine eco-tourism stands as a bridge between ocean exploration and sustainable conservation. In this chapter, we will examine how the development of eco-tourism can contribute to ocean conservation, providing economic opportunities without harming marine ecosystems.

The Unexplored Allure of the Oceans

The oceans, with their vastness and beauty, offer a unique stage for eco-tourism. From coral reefs to marine migrations, there is an underwater world of wonders waiting to be discovered. Marine eco- tourism aims to bring people closer to this beauty without compromising conservation.

Sustainable Snorkeling and Diving: Activities like snorkeling and diving allow visitors to explore marine habitats without disturbing organisms. Sustainable practices, such as "no-touch" policies and the adoption of responsible diving techniques, protect ecosystems.

Whale and Dolphin Watching: Excursions for whale and dolphin watching offer a unique and educational experience. Eco-tourism enterprises follow strict guidelines to ensure that activities do not negatively impact the behavior of marine creatures.

Low-Impact Environmental Boating Tourism: The use of low-impact environmental boats reduces the impact on the marine environment during excursions. Cleaner engines, responsible anchoring practices, and adherence to non-disturbance rules are key elements.

The Link between Eco-Tourism and Conservation

The development of marine eco-tourism is not just a business opportunity; it is also an effective means to promote ocean conservation. This link between tourism and conservation creates a virtuous circle where the beauty of the oceans becomes the engine of their protection.

Biodiversity Awareness and Appreciation: Direct contact with marine biodiversity through eco- tourism activities increases awareness and appreciation. People who develop an emotional connection with the oceans are more likely to support conservation initiatives.

Funding for Conservation: A portion of eco-tourism proceeds can be directly allocated to marine conservation projects. This funding helps support marine reserves, research programs, and local initiatives to protect marine habitats.

Environmental Education in Action: Eco-tourism businesses often integrate educational programs into their activities. Expert guides provide information on marine species, ocean conservation, and sustainable practices to promote a deeper understanding among visitors.

Successful Examples of Marine Eco-Tourism

Galápagos, Ecuador: The Galápagos Islands are an example of how eco-tourism can contribute to conservation. Wildlife observation activities are closely regulated to protect fragile ecosystems, and a portion of tourism proceeds supports conservation programs.

Great Barrier Reef, Australia: The Great Barrier Reef is one of the world's most famous marine eco- tourism destinations. Local businesses adopt sustainable practices to protect the health of the coral reef, while visitors contribute financially to its conservation.

Farne Islands, United Kingdom: The Farne Islands are known for colonies of seabirds and seals. Boat excursions are managed sustainably, ensuring that human activities do not disturb wildlife populations.

Challenges and Solutions

Excessive Tourist Influx: One of the risks of eco-tourism is the excessive influx of tourists, which can strain marine ecosystems. Limiting the number of visitors, establishing access quotas, and implementing seasonal closure periods are solutions to manage this challenge.

Impacts of Human Activities: Even well-managed eco-tourism activities can have impacts on the environment. Carefully monitoring and assessing the environmental impacts of such activities are crucial for making corrections and continuous improvements.

Involvement of Local Communities: Ensuring that local communities are involved in decisions regarding eco-tourism is fundamental. This not only ensures respect for local traditions but also provides economic opportunities for communities.

Technologies for Marine Eco-Tourism

Environmental Monitoring Applications: Environmental monitoring apps allow visitors to report and document any environmental violations or unsustainable behaviors. This encourages responsibility and transparency in the eco-tourism industry.

Underwater Observation Technologies: Advanced technologies, such as underwater cameras and remotely operated underwater vehicles (ROVs), allow visitors to explore the oceans without disturbing organisms. These technologies offer immersive experiences without negative impacts.

Sustainable Booking Systems: Online booking platforms can implement sustainability criteria in their choices of eco-tourism partners. This helps visitors make informed choices and promotes sustainable practices in the industry.

The Future of Marine Eco-Tourism

The future of marine eco-tourism is promising but requires a careful and committed approach. To ensure that this form of tourism continues to be a positive force for ocean conservation, ongoing efforts are needed.

Effective Regulations: Governments and local authorities must implement effective regulations to guide the eco-tourism industry toward sustainable practices. Collaboration between the private sector, environmental organizations, and government bodies is crucial.

Education and Awareness: Continuing to educate visitors about the importance of marine conservation and sustainable practices is crucial. Public awareness contributes to creating a demand for responsible eco-tourism.

Investments in Conservation: Increased direct investments in marine conservation through eco- tourism proceeds are essential. These funds can support research projects, the creation of marine reserves, and initiatives for reducing marine pollution.

Conclusion: A Journey Towards Conservation

Marine eco-tourism offers a fascinating view of the oceans without compromising their health. With a thoughtful, conservation-based approach and the involvement of local communities, it can become a driving force for the preservation of marine ecosystems. Each journey becomes a step towards awareness, understanding, and tangible support for the fragile beauty of the oceans. In a world where tourism is constantly growing, guiding this growth sustainably is essential to ensure that future generations can continue to explore and fall in love with the oceans as we do today.

CHAPTER 12: ENVIRONMENTAL EDUCATION – CULTIVATING LOVE FOR THE OCEANS

Environmental education emerges as a beacon in the mission to preserve the oceans. In this chapter, we will explore how to cultivate love for the oceans through innovative educational programs. The goal is to raise awareness about the importance of the oceans and inspire people to become active advocates for marine conservation.

The Power of Environmental Education

Environmental education is a powerful tool for shaping the minds and hearts of individuals. Cultivating awareness and love for the oceans from a young age is crucial for building a strong foundation of marine conservation advocates.

Emotional Connection with the Ocean: Educational programs must go beyond the transmission of information. They should create an emotional connection between students and the oceans, nurturing a love that will drive them to care for it throughout their lives.

Awareness of Critical Issues: Environmental education must address critical issues threatening the oceans, such as pollution, overfishing, and climate change. By providing a thorough understanding of these problems, the next generation is prepared to effectively address marine challenges.

Encouraging Active Action: Students should be encouraged to translate knowledge into action. Educational programs that promote practical projects, volunteer activities, and local initiatives foster active participation in marine conservation.

Innovations in Marine Environmental Education

Virtual Reality and Augmented Reality: Use technology to take students on virtual dives into the oceans. Through the use of virtual and augmented reality, they can explore marine habitats without leaving the classroom, creating an engaging and memorable experience.

Interactive Online Programs: Create interactive online educational programs that engage students through quizzes, games, and multimedia resources. These programs can be accessible from anywhere, expanding the reach of marine environmental education.

Global Collaborations: Foster collaborations between schools and educational institutions worldwide. Students can exchange information, participate in joint projects, and develop a global perspective on marine conservation.

Innovative Educational Programs

"Adopt a Wave" - Connecting with Marine Reserves: Programs that allow schools to "adopt" marine reserves. Students monitor marine life, participate in remote monitoring projects, and develop a personal connection with a specific marine environment.

Sea Festivals in Schools: Organize Sea festivals in schools, involving students in artistic, scientific, and cultural activities related to the oceans. These events celebrate the beauty of the oceans and promote awareness of their importance.

Marine Environmental Education Programs: Take students

directly to the sea for immersive learning experiences. Through field trips, beach cleanup activities, and on-site labs, students experience the oceans tangibly.

Challenges and Solutions in Marine Environmental Education Limited Access to Technology: In some communities or countries, limited access to technology can hinder participation in online educational programs. Solutions include creating printed educational materials and organizing local events.

Financial Sustainability of Programs: The financial sustainability of educational programs can be a challenge. Collaborations with environmental organizations, businesses, and governments can help ensure long-term funding.

Inclusivity and Diversity: Ensuring that educational programs are inclusive and reflect the diversity of human experiences is crucial. Representation of diverse perspectives and voices helps engage a broader audience.

Involving Families in Environmental Education

Involving families in environmental education is a crucial step to create long-term impact. Creating programs that involve parents and children promotes shared understanding and a family culture of love for the oceans.

Family Conservation Activities: Organize marine conservation activities that entire families can participate in. These activities may include beach cleanups, coastal reforestation projects, and marine life monitoring initiatives.

Stories for All Ages: Create stories and narratives about marine conservation that are suitable for all ages. These materials can be used as resources in schools and families to stimulate discussion and learning.

Family Days at the Sea: Organize family-oriented days at the sea, where parents and children can participate in educational activities, guided tours, and interactive games. These days offer a shared experience that strengthens the bond between families and oceans.

The Lasting Impact of Environmental Education

Conservation Ambassadors: Students participating in marine educational programs can become ambassadors for conservation. Spreading their knowledge and passion in their communities contributes to creating a network of ocean advocates.

Indigenous Community Initiatives: Develop indigenous community initiatives based on traditional knowledge and involve communities in teaching and preserving sustainable practices.

Student Research Projects: Encourage student research projects on marine themes. These projects not only provide valuable data for conservation but also inspire students to pursue careers in marine sciences.

Conclusions: A Sustainable Future Through Environmental Education

Environmental education is the key to building a future where the oceans thrive. Cultivating love for the oceans from a young age, involving families and communities, is essential to create a generation of aware and active citizens in marine conservation. With innovative, inclusive, and sustainable programs, we can open the door to a future where every individual feels called to protect the oceans as an essential part of our planet. Education is the beacon guiding the way to a world where the oceans, and those who love them, prosper.

CHAPTER 13: RIGHTS OF MARINE ANIMALS - A NEW ETHICAL FRONTIER

In recent decades, awareness of the importance of marine inhabitants and the need to protect their dignity has given rise to an emerging concept: the rights of marine animals. In this chapter, we will explore this new ethical frontier and discuss how we can ensure the respect and protection of marine inhabitants.

A Change in Perspective

Traditionally, marine animals have been viewed as resources to exploit rather than individuals with rights. However, with an increased understanding of the complexity of marine lives and their crucial role in ecosystems, there is a growing awareness of the need to treat marine animals with dignity and respect.

Intelligence and Social Complexity: Scientific studies have revealed the intelligence and social complexity of many marine species, such as cetaceans, octopuses, and fish. This awareness has led to a reconsideration of their ethical standing, pushing toward the recognition of their rights.

Sensitivity and Awareness: Marine animals demonstrate sensitivity and awareness of their environment. Understanding how they react to pain, loss, and human interactions has fueled the demand for ethical protection through the granting of rights.

Rights of Marine Animals

Right to Life and Liberty: The fundamental basis of the rights of marine animals is the right to life and liberty. This implies recognizing their intrinsic value as sentient beings, not merely as resources for humanity.

Prohibition of Capture and Overexploitation: The rights of marine animals include the prohibition of capture and overexploitation. This implies strict limits on commercial fishing, whaling, and other activities that jeopardize the survival and well-being of marine animals.

Protected Natural Environments: Recognizing the right of marine animals to live in protected natural environments is essential. This means creating and maintaining marine reserves where critical habitats can thrive without harmful human interference.

Challenges in Implementing the Rights of Marine Animals

Conflicting Economic Interests: The main challenge in implementing the rights of marine animals is the resistance of economic interest's dependent on marine activities, such as commercial fishing and the fishing industry. Balancing economic sustainability with the protection of marine animals requires a thoughtful approach and a gradual transition.

Gaps in Scientific Knowledge: The lack of a comprehensive understanding of the specific needs of many marine species makes it difficult to establish effective regulations. Investing in scientific research is essential to fill these gaps and inform ethically based decisions.

Impact of Human Activities: Human activities, such as pollution and climate change, continue to threaten the oceans and marine life. Ensuring the rights of marine animals requires significant actions to reduce the impact of human activities on the marine environment.

Strategies to Ensure the Rights of Marine Animals

International Legislative Reform: Create and strengthen international laws that protect marine animals and ensure respect for their rights. Binding global treaties can establish a normative foundation for the ethical protection of marine inhabitants.

Public Education on the Dignity of Marine Animals: Create educational programs that raise public awareness about the dignity of marine animals. A cultural shift in which society recognizes and respects marine animals as individuals deserving of protection is crucial for the success of such efforts.

Promote Sustainable Alternatives: Advocate for the development and adoption of sustainable alternatives to activities that threaten marine animals. This could include selective fishing technologies, less invasive fishing practices, and sustainable alternative food sources.

Key Sectors in Implementing the Rights of Marine Animals

Sustainable Fishing: Reform fishing practices to ensure sustainability and the protection of marine animals. This includes reducing overfishing, adopting selective fishing technologies, and creating marine protected areas.

Responsible Tourism: Promote responsible tourism that respects marine habitats and the creatures that inhabit them. Regulate marine animal-watching activities to avoid disturbances and harm to individuals and ecosystems.

Conservation of Marine Habitats: Focus on creating and managing marine reserves that provide safe and protected environments for marine animals. This can help preserve biodiversity and ensure that marine ecosystems remain intact.

Case Study: The Declaration of Rights of Marine Animals

Imagine a global declaration that clearly establishes the rights of marine animals. This document could include:

Recognition of Sensitivity and Intelligence: Explicitly affirm the recognition of the sensitivity and intelligence of marine animals based on scientific evidence.

Ban Harmful Practices: Prohibit practices such as capture and overexploitation, as well as any form of unethical mistreatment or capture.

Promote Conservation and Rehabilitation: Encourage the conservation of marine habitats and promote rehabilitation programs for injured or mistreated individuals.

Responsibility of Future Generations: Emphasize the responsibility of future generations in protecting marine animals and managing the oceans sustainably.

Conclusion: An Ethical Future for Marine Inhabitants

Ensuring the rights of marine animals is a complex challenge, but it is essential to create an ethical future for marine inhabitants. As our understanding of marine life grows, it is our duty to evolve our practices and policies to reflect this awareness. Only through ethical and committed actions can we hope to create a world where marine animals live without fear of exploitation and in harmony with oceanic ecosystems. This new ethical frontier is a call to action to protect and respect the individuals who share our planet and the oceans we all inhabit.

CHAPTER 14:
BLUE ECONOMY – SUSTAINABILITY AND PROSPERITY

In the modern era, the idea of economic prosperity can no longer ignore the importance of preserving the oceans. The concept of the "Blue Economy" emerges as an innovative approach, embracing economic development in harmony with ocean conservation. In this chapter, we will explore the concept of the Blue Economy and how it can become a driver for long-term sustainability and prosperity.

Foundations of the Blue Economy

The Blue Economy represents a crucial transition from the traditional view of the ocean as an unlimited exploitable resource to a perspective that recognizes the need to balance economic development with the health of the oceans. These are its foundations:

Conservation and Sustainable Use: The Blue Economy is based on the conservation of oceans and the sustainable use of marine resources. This implies thoughtful management of human activities involving oceans, such as fishing, tourism, and maritime industries.

Technological Innovation: It incorporates technological innovation to maximize efficiency and reduce the environmental impact of oceanic activities. Advanced technologies such as ocean sensors, underwater drones, and satellite monitoring are crucial tools for intelligent ocean management.

Engagement of Local Communities: It promotes the active engagement of local communities in decisions regarding the oceans. This bottom-up approach ensures that solutions are tailored to the specific needs of each region.

Key Sectors of the Blue Economy

Sustainable Fishing: It transforms the fishing sector from a source of exploitation to a driver of sustainability. Sustainable fishing involves strict limits, selective fishing techniques, and the creation of marine reserves to allow for the regeneration of fish populations.

Responsible Tourism: It develops tourism as an economic engine that respects marine ecosystems. Responsible tourism involves regulations to avoid negative environmental impacts, such as marine animal watching and well-managed ecotourism.

Blue Energy: It explores renewable energy sources from the oceans, such as marine current energy, wave energy, and ocean thermal energy. These sources reduce dependence on unsustainable energy sources and mitigate environmental impacts.

Sustainable Bioprospecting: It examines marine biological resources sustainably to develop medicines, materials, and other products. This sector requires ethical regulations to ensure that bioprospecting does not endanger marine biodiversity.

Benefits of the Blue Economy

Conservation of Biodiversity: The Blue Economy is a bulwark for marine biodiversity. Through sustainable management practices, the variety of life in the oceans is preserved, contributing to the health of ecosystems.

Sustainable Economic Growth: It integrates economic growth with sustainability, promoting long- term prosperity. Economic activities related to the oceans become catalysts for sustainable development rather than environmental threats.

Job Creation: The Blue Economy creates job opportunities in coastal communities. Sectors such as sustainable fishing, responsible tourism, and blue energy generate local employment, contributing to the economic resilience of coastal regions.

Technological Innovation and Scientific Research: It stimulates technological innovation and scientific research. The exploration and sustainable management of the oceans require advances in technology and a deep understanding of marine ecosystems.

Challenges to Address in the Blue Economy

Resistance to Change: Addressing resistance to innovative approaches in industry and governance. Many sectors are entrenched in traditional practices and may resist necessary changes to embrace the Blue Economy.

Need for Long-Term Investments: The Blue Economy requires long-term investments, which could be challenging in contexts where short-term vision dominates economic decisions.

Governance and International Coordination: Coordinating efforts globally. Oceans know no national boundaries, so effective governance and international cooperation are crucial to ensuring ocean sustainability.

Tools and Policies for the Blue Economy

Sustainability Certifications: Implementing sustainability certifications for sectors such as fishing and tourism. These certificates inform consumers about sustainable practices and incentivize businesses to adhere to high standards.

Marine Protected Areas: Creating and managing marine

protected areas. These areas serve as reserves where biodiversity can thrive without the pressures of harmful human activities.

Tax Incentives for Sustainable Practices: Offering tax incentives to businesses engaged in sustainable practices related to the oceans. These incentives can promote the transition to a blue economy.

Case Study: The Blue Economy Model in the Maldives

The Maldives is an excellent example of how a country can successfully adopt the Blue Economy. Through measures such as the creation of marine protected areas, the adoption of sustainable fishing practices, and the development of responsible tourism, the Maldives is balancing economic development with ocean conservation. This model could inspire other countries to follow a similar path.

Conclusion: A Future of Sustainability and Prosperity

The Blue Economy offers a promising path to integrate economic development with ocean conservation. Embracing this approach not only preserves the richness of marine ecosystems but also creates opportunities for sustainable economic growth. However, to fully realize the benefits of the Blue Economy, it is essential to address challenges and adopt policies and practices that balance human prosperity with the health of the oceans. Only through global commitment and thoughtful governance can we shape a future where the oceans are a source of sustainability and prosperity for generations to come. The Blue Economy is not just a concept; it is an invitation to create a future where our prosperity is intricately intertwined with the health of our oceans.

CHAPTER 15: ENVIRONMENTAL LOBBYING - VOICE OF THE OCEANS WORLDWIDE

In the vast landscape of ocean conservation, environmental lobbies emerge as powerful guardians, dedicated to preserving marine life and ensuring the sustainability of oceanic ecosystems. In this chapter, we will examine the crucial role of environmental organizations as the voice of the oceans worldwide and provide suggestions on how to enhance the effectiveness of environmental lobbying globally.

The Need for Environmental Lobbying for the Oceans

Oceans, often overlooked in political and economic decisions, require a strong and dedicated voice to ensure their protection and sustainable management. Environmental lobbies play a fundamental role in this context for various reasons:

Awareness and Information: Environmental lobbies contribute to raising awareness among the public and decision-makers about the importance of oceans. They inform about crucial issues such as pollution, overfishing, and climate change, advocating for the adoption of more sustainable policies.

Representation of Ocean Interests: Environmental lobbies act as dedicated advocates for ocean interests, often countering short-term economic interests that threaten the health of marine ecosystems.

Monitoring Policies and International Agreements: They play a critical monitoring role on policies and international agreements concerning oceans. This ensures that globally made decisions are oriented towards sustainability and conservation.

Promotion of Innovative Solutions: Environmental lobbies promote innovative solutions to address threats to the oceans. These solutions range from sustainable fishing to renewable marine energies, contributing to shaping a more sustainable future for the oceans.

Challenges Faced by Oceanic Environmental Lobbies

Limited Funding: Many environmental lobbies operate with limited financial resources, making it challenging to compete with well-funded economic interests. This can affect their ability to conduct effective campaigns and support long-term projects.

Resistance to Change: Addressing resistance to innovative approaches from established economic sectors. Environmental lobbies often need to overcome resistance to new policies and practices that might threaten short-term interests.

Difficulty in International Cooperation: Coordinating efforts globally can be challenging, especially considering the diverse priorities and interests of countries. International cooperation is essential to effectively address oceanic issues.

Conflicts of Interest and Political Influences: Environmental lobbies must navigate conflicts of interest and political influences that can undermine their efforts. The influence of industrial lobbies can distort decision-making processes in favor of practices harmful to the oceans.

Strategies to Improve the Effectiveness of Oceanic Environmental Lobbies

Diversification of Funding Sources: Environmental lobbies can seek to diversify their sources of funding. In addition to public donations, they can explore partnerships with foundations, ethical companies, and patrons interested in ocean conservation. Collaboration and Networking: Creating networks and collaborations among different environmental lobbies, non-governmental organizations (NGOs), and research groups. Collective strength can increase impact and provide a broader platform for ocean advocacy.

Effective Communication: Investing in effective communication strategies to engage the public and policymakers. Engaging narratives, clear scientific data, and success stories can generate interest and support for the cause.

Technological Innovation in Advocacy: Leveraging innovative technologies in advocacy. Using digital platforms, social media, and data visualization tools to reach a wider audience and raise awareness about oceanic issues.

Successes of Oceanic Environmental Lobbies

Ban on Single-Use Plastic: In many countries, environmental lobbies have contributed to promoting legislation banning or limiting the use of single-use plastics. These bans aim to reduce plastic pollution in the oceans.

Creation of Marine Protected Areas: Environmental lobbies have successfully supported the creation of marine protected areas in various parts of the world. These areas provide refuge for threatened marine ecosystems and promote biodiversity conservation.

Awareness Campaigns on Endangered Species: Environmental lobbies often lead awareness campaigns on endangered marine species, such as sea turtles, whales, and dolphins. These campaigns aim to protect critical habitats and reduce human threats to these species.

Case Study: Greenpeace and the "Save the Oceans" Campaign

Greenpeace, one of the most prominent environmental organizations, has conducted significant campaigns for ocean conservation. Their "Save the Oceans" campaign has garnered global attention on marine threats, advocating for stricter policies on sustainable fishing and the reduction of plastic pollution.

The Future of Oceanic Environmental Lobbies

The future of oceanic environmental lobbies requires constant evolution and adaptation to emerging challenges. Some key directions for the future include:

Integration of Local Communities: Actively involving local communities in ocean conservation initiatives, recognizing their local knowledge and crucial role in conservation.

Combatting Climate Change: Focusing on advocacy for policies and actions that address climate change, one of the most severe threats to the oceans.

Technological Monitoring: Harnessing advanced technologies such as satellite monitoring and artificial intelligence to monitor and document threats to the oceans in real-time.

Continuous Environmental Education: Continuing to invest in educational programs to keep the public informed and engaged in ocean conservation.

Conclusion: A Global Voice for the Oceans

Environmental lobbies are the global voice of the oceans, striving to ensure a sustainable future for marine ecosystems. While facing challenges and resistance, their unwavering commitment is essential to ensure that the oceans are preserved for future generations. By enhancing the effectiveness of environmental lobbies and promoting greater awareness, we can hope to shape a future where the oceans are treated with the dignity and care they deserve. Ocean conservation is a global cause, and environmental lobbies are the voice echoing worldwide to protect these precious ecosystems.

CHAPTER 16: LAWS AND POLICIES - FOUNDATIONS FOR OCEAN-RESPONSIBILITY

The oceans, with their vastness and complexity, require a robust legal framework and effective policies to ensure their sustainable management and the conservation of their biodiversity. In this chapter, we will examine existing laws and policies related to the oceans, highlighting current challenges, and propose innovative measures to strengthen international ocean governance.

The Current State of Oceanic Laws and Policies

United Nations Convention on the Law of the Sea (UNCLOS)

The United Nations Convention on the Law of the Sea (UNCLOS) represents the primary international legal instrument for the oceans. Adopted in 1982, UNCLOS establishes the legal framework for the peaceful use of the oceans, the management of marine resources, and the conservation of the marine environment. However, despite its successes, some challenges persist:

Enforcement Issues: Despite being ratified by most countries, the uniform enforcement of UNCLOS remains a challenge. Some countries do not fully comply with its provisions, leading to uneven ocean governance.

Lack of Effective Sanction Mechanisms: UNCLOS lacks effective sanction mechanisms to address violations and harmful behavior. The lack of significant consequences can weaken the treaty's effectiveness.

Regional and National Agreements

In addition to UNCLOS, numerous regional and national agreements aim to address specific oceanic issues. However, the fragmentation of efforts can lead to inconsistencies and gaps in ocean protection.

National and Local Policies

National and local policies vary widely in terms of approaches to ocean management. Some countries adopt progressive policies, promoting sustainability and conservation, while others may be slower to respond to oceanic challenges.

Challenges in Current Ocean Governance

Transboundary Impacts: Oceans do not recognize national borders, and the actions of one country can have significant impacts on others. Lack of internation cordination can make it difficult to address issues such as pollution and overfishing effectively.

Lack of Fair Representation: Developing nations often have fewer resources to participate in international negotiations and implement advanced ocean policies. This can lead to a lack of fair representation in global decision-making processes.

Need for Adaptation to New Threats: Current laws and policies often do not adequately address emerging threats, such as climate change and the growth of aquaculture. Constant adaptation is necessary to tackle evolving challenges.

Lack of Adequate Financial Mechanisms: The lack of adequate funding limits the ability of many countries to implement sustainable ocean policies. The creation of innovative financial mechanisms is essential to support long-term initiatives.

Innovative Proposals for Strong Ocean Governance

Creation of an International Ocean Organization

Proposal: Establish an international organization dedicated exclusively to ocean management. This organization would be responsible for coordinating global efforts, monitoring the enforcement of existing ocean laws, and developing new regulations when necessary.

Benefits:

Effective international coordination.

Provides a platform to address

emerging challenges. Improves

fair representation of all nations.

Global Fund for Ocean Conservation

Proposal: Create a global fund dedicated to ocean conservation. This fund would be financed through voluntary contributions from countries and international organizations and used to fund conservation projects, research, and the enforcement of ocean laws.

Benefits:

Provides consistent financial resources

for ocean initiatives. Promotes

international cooperation through shared

funding. Addresses the lack of financial

resources for ocean management.

Ecosystem-Based Approach to Marine

Resource Management

Proposal: Adopt an ecosystem-based approach to marine

resource management, considering the oceans as interconnected systems. This means shifting focus from managing individual species to conserving marine ecosystems as a whole.

Benefits:

Promotes the conservation of

marine biodiversity. Reduces the

risk of bycatch and habitat

damage.

Contributes to maintaining the ecological

balance of the oceans. Development of

Advanced Monitoring Technologies

Proposal: Invest in the development and implementation of advanced ocean monitoring technologies, such as satellites, ocean sensors, and underwater drones. These technologies would enable constant surveillance of the oceans, facilitating real-time data collection and a rapid response to emergencies.

Benefits:

Improves the ability to monitor and

respond to ocean threats. Provides

accurate data to assess policy

effectiveness.

Supports scientific

research on ocean health.

Tax Incentives for Ocean

Sustainability

Proposal: Introduce tax incentives for businesses and nations adopting sustainable practices related to the oceans. This could include tax breaks for sustainable fishing, the adoption of renewable marine energies, and the reduction of environmental impact from maritime activities.

Benefits:

Promotes the transition to more sustainable practices. Stimulates innovation and the adoption of green technologies.

Creates an economic environment supportive of ocean conservation.

Implementation and Future Challenges

Implementing innovative proposals will require global collaboration and ongoing commitment. Challenges will be inevitable, but addressing them is crucial to ensuring ocean sustainability. The adoption of a strong and flexible ocean governance is essential to respond to emerging challenges and protect the oceans for future generations.

Conclusion: An Ocean-Responsible Future

Ocean-responsible management requires joint commitment and concrete actions at the international level. Through the creation of new laws and innovative policies, we can shape a future where the oceans are preserved as precious resources and protected as unique ecosystems. Only with strong and collaborative ocean governance can we hope to preserve the beauty, diversity, and vitality of the oceans for generations to come.

CHAPTER 17: GLOBAL COLLABORATION - JOINING FORCES FOR THE OCEANS

Due to their vast and interconnected nature, oceans require unprecedented global collaboration to be adequately preserved and sustainably managed. In this chapter, we will explore the importance of international collaboration in ocean conservation and propose new models of cooperation to address global challenges related to the oceans.

The Case for Global

Collaboration Oceans

Without Borders

Oceans do not recognize national borders. This means that the actions of one country can have significant impacts on others. Pollution, overfishing, and climate change are global threats that require coordinated international solutions.

Global Biodiversity

Marine biodiversity is a global heritage. The migrations of marine species span oceans and involve different nations. Protecting these species and their habitats requires joint efforts to ensure that the actions of one country do not compromise the conservation efforts of another.

Climate Change and Ocean Acidification

Climate change and ocean acidification are global challenges that require a coordinated response. The impacts of these threats extend well beyond national borders, and only through global collaboration can we develop effective strategies for mitigation and adaptation.

Equitable Access to Marine Resources

Access to marine resources, such as fishing, must be managed fairly and sustainably. International collaboration is essential to develop agreements that balance the needs of nations and ensure the conservation of fishery resources globally.

Models of Cooperation for

Ocean Conservation Global

Action Plan for Ocean Plastic

Proposal: Create a global action plan to address the issue of plastic in the oceans. This plan would include binding commitments from countries to reduce the use of single-use plastic, implement effective recycling systems, and combat plastic pollution.

Benefits:

Significant reduction in global plastic

pollution. Standardization of plastic

management practices worldwide.

Collaboration among industries, governments, and non-governmental organizations (NGOs) to tackle the challenge.

Global Network of Marine Protected Areas

Proposal: Create a global network of marine protected areas, involving nations worldwide in the conservation of critical marine habitats. This network would facilitate the migration of marine species, protecting biodiversity and promoting sustainable management of marine resources.

Benefits:

Conservation of marine biodiversity on a global scale.

Promotion of sustainable fishing through the creation of safe zones for the reproduction and growth of marine species.

Collaboration among nations for the shared

management of marine areas. International

Fund for Ocean Conservation

Proposal: Create an international fund exclusively dedicated to ocean conservation. This fund would be financed by voluntary contributions from nations, international organizations, and the private sector and would be used to fund research, monitoring, and marine conservation projects.

Benefits:

Consistent financial resources for ocean conservation

and management. Collaboration between the public

and private sectors to address oceanic challenges.

Promotion of international cooperation through

shared funding.

Collaboration in Ocean Research

Proposal: Foster international collaboration in ocean research. Establish joint research programs and data exchange between nations to improve understanding of the impacts of climate change on the oceans, monitor marine biodiversity, and develop innovative solutions for ocean challenges.

Benefits:

Expansion of global knowledge

about the oceans. Identification of

common solutions to ocean challenges.

Promotion of technology and expertise

sharing among nations. Cooperation for

Maritime Security

Proposal: Increase international cooperation to ensure maritime security. This would include sharing information on illegal activities at sea, collaborating in the fight against pirate fishing, and promoting global standards for safe navigation.

Benefits:

Reduction of illegal activities at sea

through coordinated actions. Protection

of marine resources from harmful

practices.

Creation of a safe maritime

environment for all nations.

Addressing Challenges of Global

Collaboration Resilience to

Geopolitical Tensions

Geopolitical tensions can complicate international cooperation. To address this challenge, it is essential to promote dialogue and diplomacy as means to resolve disputes and encourage collaboration for ocean conservation.

Promoting Equality and Inclusivity

Ensuring that all nations, regardless of their level of development, have a voice in ocean decisions is crucial. Promoting policies that favor equality and inclusivity can reduce disparities in participation in international cooperation.

Long-Term Financial Sustainability

Ensuring the financial sustainability of ocean conservation initiatives requires the creation of innovative financial mechanisms. This can include introducing taxes on activities harmful to the oceans and establishing financial incentives for sustainable practices.

Global Education and Awareness

Promoting global awareness of the importance of oceans is fundamental to ensuring public support for international cooperation. Educational programs and awareness campaigns can contribute to informing and engaging the public in the ocean conservation process.

Conclusion: A Future of Collaboration for the Oceans

Global collaboration is the key to addressing the complex and interconnected challenges related to the oceans. Through innovative models of cooperation, we can hope to create a future where the oceans are preserved as precious resources for generations to come. By joining forces internationally, we can shape an ocean-responsible destiny and ensure that the wonders of the oceans are safeguarded forever.

CHAPTER 18: THE FUTURE OF THE OCEANS - A CALL TO ACTION

In this concluding chapter, let's delve into the hopes and challenges that shape the future of the oceans. It is a call to action, an invitation to actively participate in the safeguarding of this global treasure, so it may continue to thrive and inspire future generations.

Hopes for the Future

of the Oceans Growing

Global Awareness

A bright hope for the future of the oceans is the increasing global awareness of the importance of marine conservation. With the spread of information and educational initiatives, more people are understanding the crucial role oceans play in climate regulation, oxygen production, and the provision of food resources.

Revolutionary Technological Innovations

The future of the oceans could be shaped by revolutionary technological innovations. Ocean drones, advanced sensors, and artificial intelligence can contribute to real-time ocean monitoring, identify issues early on, and drive technology-based solutions to address ocean challenges.

Global Commitment to Sustainability

A key hope is the increase in global commitment to sustainability. Governments, international organizations, the private sector, and individuals must collaborate to develop and implement sustainable policies, resource management practices, and initiatives that ensure the long-term health of the oceans.

Conservation of Marine Biodiversity

Conservation of marine biodiversity is essential for the future of the oceans. We hope that targeted efforts to create marine protected areas, sustainable fisheries management, and the protection of critical habitats can preserve the unique diversity of life in the oceans.

Growth of Responsible Eco-Tourism

Responsible eco-tourism could become a driver for ocean conservation. If managed sustainably, it can offer economic opportunities for coastal communities without harming marine ecosystems, while promoting environmental awareness among visitors.

Challenges for the

Future of the Oceans

Unstoppable Climate

Change

One of the most significant challenges is posed by unstoppable climate change. Rising ocean temperatures, acidification, and extreme weather events threaten marine ecosystems. Addressing this challenge will require global efforts to mitigate the impact and adapt to inevitable changes.

Increasing Anthropogenic Pressure

Population growth and the rise of anthropogenic activities

exert ever-increasing pressure on the oceans. Overfishing, pollution, habitat destruction, and the exploitation of marine resources are ongoing threats that require immediate action to reverse the trend.

Limited Financial Resources

Ocean conservation demands significant financial resources. However, the limited availability of funding can be a barrier to implementing long-term sustainable policies and initiatives. Finding innovative solutions to ensure adequate funding is crucial.

Lack of Efficient Global Coordination

The lack of efficient global coordination is a challenge in addressing ocean issues on a worldwide scale. The diversity of national interests, the absence of a dedicated international body for the oceans, and geopolitical tensions can hinder effective collaboration.

Changes in Economic Models

Addressing ocean challenges will require changes in current economic models. The adoption of sustainable practices may necessitate a transformation of industries related to the oceans, creating resistance from established sectors. Striking a balance between economic prosperity and conservation is a complex challenge.

The Call to Action

The future of the oceans is filled with uncertainty, but it is also permeated with opportunities for positive change. The call to action is directed at each individual, community, nation, and organization. Here are some paths of action that can contribute to shaping a sustainable future for the oceans:

Adopting Sustainable Lifestyle Practices

Every individual can contribute by adopting sustainable lifestyle practices. Reducing the use of single-use plastics, supporting sustainable products, saving energy, and endorsing conservation initiatives are tangible ways to make a difference.

Participating in Local Conservation Initiatives

Joining local conservation initiatives is crucial. Participating in beach clean-ups, supporting marine reforestation projects, and engaging in biodiversity monitoring programs directly contribute to the health of the oceans.

Supporting Ocean Conservation Organizations

Non-governmental organizations (NGOs) and conservation organizations play a key role in defending the oceans. Financially or voluntarily supporting these organizations can contribute to the implementation of global conservation projects.

Promoting Research and Innovation

Scientific research and innovation are fundamental to addressing ocean challenges. Supporting marine research institutes, promoting research on the sustainability of marine resources, and encouraging technological innovation can pave the way for ocean conservation.

Engaging Local Communities in Resource Management

Actively involving local communities in marine resource management is essential. Participatory management models, considering local knowledge and promoting shared responsibility, can ensure sustainable management of oceanic resources.

Conclusion: A Shared Future for the Oceans

The future of the oceans is a work in progress, shaped by the actions of anyone with an interest in the health of our planet. It is a shared responsibility that requires commitment, awareness, and action. Only by joining forces at individual and global levels can we hope to ensure that the oceans continue to be the vital heart of the Earth, pulsating with life, beauty, and diversity for generations to come. The call to action has been issued. Each one of us is called to answer the oceans' call.

EPILOGUE

As these pages come to a close, I invite you to reflect on the journey we have taken together through "Oceans at Risk." The epilogue is the point where the past and present merge, where challenges meet solutions, and where hope intertwines with responsibility.

Reflections on the Journey

In this epilogue, I invite you to look back and contemplate the oceanic wonders we have explored. Each word has been a lens through which we have examined unknown worlds, admired the diversity of marine life, and connected with the very essence of the oceans. I hope you have felt the call of the waves, listened to the stories of marine inhabitants, and learned to love and respect this vital heart of the Earth.

Assessment of Faced Challenges

Throughout the book, we faced the challenges weighing on the oceans with open eyes. The epilogue is the moment to assess the revealed threats, to recognize the complexity of the challenges the oceans face every day. We delved into pollution, overfishing, pirate fishing, climate change, and other threats, outlining the extent of these problems and understanding their interconnectedness.

Hopes for an Ocean-Responsible Future

However, the epilogue is also where we catch a glimpse of hopes for an ocean-responsible future. We explored innovative solutions, sustainable models, and strategies that can lead to positive change. Each proposal is a ray of light illuminating the path toward ocean-responsible and sustainable management. I hope these ideas have inspired a sense of confidence that we can shape a future where the oceans thrive.

The Call to Action

The epilogue is also an invitation to action. Every reader is called to translate words into tangible deeds. Whether it's reducing plastic use, supporting environmental organizations, promoting awareness, or adopting sustainable fishing practices, each of us can contribute to positive change. The invitation is to become active participants in ocean conservation, to be an integral part of the solution.

Special Thanks

In this epilogue, I want to express deep gratitude to all those who joined this journey. To you, curious readers, who dedicated time and attention to explore the depths of the oceans with me. To all the scientists, activists, environmental practitioners, and ocean guardians working tirelessly for marine conservation, your dedication is the guiding light on this path.

The Future of the Oceans

Finally, the epilogue is where the past meets the future. Looking ahead, I hope the oceans continue to be the vital heart of the Earth. May we all learn to live in harmony with the oceans, to cherish them with love and respect for generations to come. Every step we take today will shape the destiny of the oceans tomorrow.

With gratitude for your journey together and the hope that these words are waves that continue to resonate in your hearts,
Zahra Jonsson

AFTERWORD

As I set the pen down and close the book, I find myself in a space of reflection and gratitude. The afterword is where the created connects with the creator, where the author and the reader embrace through the written pages. In "Oceans at Risk," this embrace is intimate and global, reaching out to all who have shared this journey with me.

A Shared Journey

The afterword is the occasion to express my deep gratitude to you, readers. Thank you for accompanying me through the depths of the oceans, for sharing your curiosity, attention, and commitment. This book would be nothing without your virtual presence, without your eyes dancing between the lines.

Connections with the Oceans

In this afterword, let's reflect together on the connections we have forged with the oceans. Every word, every page, has been an attempt to weave a deeper connection with these unknown yet fundamental realms. I hope you have heard the roar of the sea in the words, sensed the complexity, and delicacy of marine ecosystems.

Challenges and Hopes

Looking towards the future, let's examine the challenges we have unveiled and the hopes we have glimpsed. The afterword is the moment where we embrace the complexity of reality, where we acknowledge that marine conservation is a collective mission requiring commitment, awareness, and action. Challenges are real, but solutions are possible through collaboration and dedication.

An Appeal for Persistence

The afterword is also an appeal for persistence. Oceans are resilient, but our responsibility towards them is enduring. I encourage you to carry the stories of the oceans with you, to share them with others, and to become ambassadors for marine conservation. Every small gesture counts, and together we can shape a sustainable future for the oceans.

Beyond the Pages

Finally, the afterword is the bridge to what lies beyond the pages. This book is only the beginning of the journey. I invite every reader to explore further, to seek knowledge, to engage in conservation initiatives, and to bring ocean awareness into their daily lives. The oceans are alive, and our commitment can make a difference.

Thank You

Thank you, from the depths of my heart, to all who made this journey possible. To the researchers and experts who shared their knowledge, to the readers who brought these words into their homes, and to the waves themselves that inspired us. May your lives always be enveloped by the wonder of the oceans.

With gratitude and hope for our ocean-responsible future,

Zahra Jonsson

ACKNOWLEDGEMENT

Dear readers,

As I sit down to write these acknowledgments, I feel the need to express deep gratitude for every single reader who has journeyed through the pages of "Oceans at Risk." This book is the result of a shared voyage, and your virtual presence has made this journey extraordinary.

Thank you for your curiosity, for opening this book with open hearts and minds ready for exploration. The fact that you have dedicated your time and attention to explore the wonders and challenges of the oceans means everything to me. I hope every word has resonated with you, transporting you to new worlds and inspiring profound reflections.

I thank those who have shared their stories, experiences, and knowledge about marine conservation. Each contribution has enriched the fabric of this book, providing unique and valuable perspectives beyond my words.

A special thanks go to scientists, activists, environmentalists, and all those tirelessly working for ocean conservation. Your voices are beacons of hope, and your dedication is a source of inspiration for all of us.

I also thank those who, through these pages, have committed to becoming active participants in ocean conservation. Every action, big or small, matters, and your willingness to make a difference is a guiding light toward an ocean-responsible future.

Finally, I want to thank the waves themselves. The oceans are infinite sources of inspiration, wisdom, and beauty. Thank you for being the muse of this book and for continuing to narrate stories of life, hope, and wonder.

With deep gratitude and affection,
Zahra Jonsson

ABOUT THE AUTHOR

Zahra Jonsson

Zahra Jonsson is a prominent figure in the landscape of environmentalism dedicated to protecting our precious planet. Her passion for environmental conservation encompasses a wide range of issues, with a particular focus on the ocean, forests, wildlife, and the environment as a whole.

Her journey in environmentalism has been shaped by the intrinsic belief that the Earth is an interconnected ecosystem, and protecting every part of it is crucial to ensuring a sustainable future. Zahra has emerged as a guiding force in safeguarding the oceans, tirelessly working to preserve marine life and counter environmental threats weighing on the waters.

In addition to her commitment to the oceans, Zahra is a staunch advocate for forest conservation. She has actively worked towards the protection of forests, recognizing the vital role they play in mitigating climate change and preserving biodiversity.

Her love for animals is evident in her commitment to protecting wildlife from any form of threat. Zahra advocates for ensuring that vulnerable species are preserved, promoting awareness of the importance of respecting and coexisting with all creatures that share our planet.

Zahra Jonsson also stands out for her resolute rejection of nuclear energy, emphasizing the importance of sustainable energy sources to preserve the health of our environment and reduce the negative impact on the global ecosystem.

Her approach to environmentalism is inclusive and interdisciplinary, seeking to address environmental challenges in a holistic way. Zahra is involved in educational initiatives, awareness campaigns, and direct actions to inspire positive changes in collective mindset and environmental policies.

Through her multifaceted commitment, Zahra Jonsson positions herself as an authoritative voice in contemporary environmentalism, demonstrating that concrete action, widespread awareness, and passion can shape a future where the Earth and all its inhabitants thrive.

BOOKS BY THIS AUTHOR

Save The Planet: Climate Change And The Fight For Environmental Sustainability.

"Save the Planet: Climate Change and the Fight for Environmental Sustainability" is a comprehensive and enlightening book that delves into the urgent global issue of climate change and provides actionable insights on how we can collectively work towards a sustainable future. Authored by Zahra Jonsson, a renowned environmentalist and advocate for sustainable living, this book offers a compelling exploration of the challenges posed by climate change and the measures we can take to mitigate its impacts.

In this thought-provoking book, Zahra Jonsson takes readers on a transformative journey, unraveling the complex web of climate change causes, consequences, and potential solutions. Drawing upon the latest scientific research, historical context, and real-life examples, Jonsson presents a compelling case for urgent action and the need to prioritize environmental sustainability.

The book consists of 16 informative and engaging chapters that cover a wide range of topics. From understanding the basics of climate change to exploring the fragility of ecosystems, measuring our carbon footprint, embracing clean energy, and examining the role of governments, businesses, and individuals in climate action, each chapter provides valuable insights and practical guidance.

With a focus on solutions, "Save the Planet" showcases innovative technologies, sustainable practices, and successful initiatives that offer hope and inspiration. The book explores topics such as sustainable agriculture, responsible consumption, water conservation, biodiversity conservation, climate justice, and the power of education and youth activism.

What sets this book apart is its balanced and holistic approach to addressing climate change. It goes beyond raising awareness and explores the interconnectedness of environmental, social, and economic factors. It emphasizes the importance of collective action, global collaboration, and the need to create a more just and sustainable world for all.

Throughout the pages of "Save the Planet," Zahra Jonsson skillfully blends scientific expertise with accessible language, making complex concepts easily understandable for readers of all backgrounds. The book is enriched with compelling anecdotes, case studies, and thought-provoking questions, encouraging readers to reflect on their own roles in the fight against climate change.

Whether you are an environmental enthusiast, a concerned citizen, a policymaker, or a student seeking a comprehensive understanding of climate change and the path to sustainability, "Save the Planet" is a must-read. Zahra Jonsson's passionate and well-researched writing will empower and equip readers with the knowledge and inspiration to take meaningful action towards a more sustainable future.

Join the movement to save the planet. Get your copy of "Save the Planet: Climate Change and the Fight for Environmental Sustainability" today and be part of the global effort to combat climate change and protect our planet for future generations.

No To Nuclear: Exploring The Downsides Of

Nuclear Power And Sustainable Alternatives For A Safe And Clean Energy Future

"NO to Nuclear" is a comprehensive work that explores the world of nuclear energy, highlighting its negative aspects and offering sustainable solutions for a safe and clean energy future. Written by Zahra Jonsson, an expert in the field of energy and the environment, this book offers an in-depth view of the issues associated with nuclear power and presents valid and promising alternatives.

Through painstaking research and detailed analysis, the author examines the risks and accidents associated with nuclear energy, such as Chernobyl and Fukushima, revealing the devastating consequences that can occur when things go wrong. The implications for the environment and human health are explored, allowing readers to fully understand the long-term impacts of such incidents.

However, the book doesn't just highlight the risks. Zahra Jonsson provides an extensive overview of alternative renewable energy sources, such as solar, wind, hydroelectric and geothermal energy. The potential of these sustainable energy sources is explored, showing how they can contribute to a secure and environmentally sustainable energy future.

The author also addresses crucial issues such as nuclear proliferation and radioactive waste disposal, providing insights into the challenges associated with these issues and presenting solutions proposed by the scientific community and environmental activists.

In addition, "NO to Nuclear" examines the role of nuclear energy in the context of climate change, analyzing the associated carbon emissions controversies and offering a balanced view of the pros and cons. The author explores the potential impact of anti-nuclear policies on national security, focusing on the complex dynamics between energy and international politics.

Concluding, Zahra Jonsson presents a comprehensive picture of nuclear energy, highlighting its negative aspects and the challenges associated with it, but also opening the door to a nuclear-free future through the adoption of sustainable and innovative energy sources. "NO to Nuclear" is an essential book for anyone interested in sustainability, clean energy and creating a better world for future generations.